Gaia and Philosophy

Gaia and Philosophy

DORION SAGAN AND LYNN MARGULIS

First published by Ignota in 2023
ignota.org

'Gaia and Philosophy' by Dorion Sagan and Lynn Margulis
was first published in *On Nature*, 1984
Reprinted by permission of Dorion Sagan

ISBN-13: 9781838003968

Design by Cecilia Serafini
Typeset in Adobe Caslon Pro by Marsha Swan
Printed and bound in Great Britain by TJ Books Limited

1 3 5 7 9 10 8 6 4 2

MIX
Paper from
responsible sources
FSC® C013056
www.fsc.org

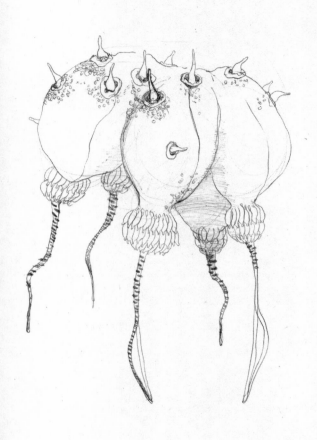

Contents

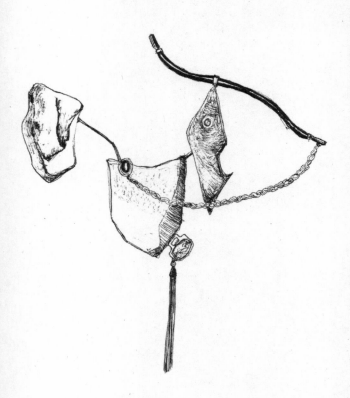

Introduction

A Body in the Form of a Planet:
Gaia, Fifty Years Later

BY DORION SAGAN

Gaia is a body in the form of a planet. Gaia describes a living Earth, an idea with precedents in natural science and philosophy for 2,500 years, and longer in many indigenous belief systems. The basis of the Gaia hypothesis is that Earth's atmosphere is as complicated as the blood or skin of an animal. It was formally proposed by chemist James E. Lovelock (1919–2022) in 1972, and its name was provided by his neighbour, William Golding, author of *Lord of the Flies*. Lovelock conjectured that Earth's surprisingly metastable chemistry was maintained by life's activity. Other environmental variables, such as global mean temperature and marine salinity, and even the continuing presence of water on the planet, are theorised to be partly dependent on the activity of live organisms, effectively making the entire planetary surface alive. The evolutionary biologist Lynn Margulis (1938–2011), my mother and long-time writing partner, helped turn Lovelock's hypothesis into a full-fledged theory: she posited that specifically gas-exchanging microbial life, which has been present on Earth's surface for billions of years, is responsible for giving our planet the character of a living body.

Margulis's former husband, Carl Sagan (1934–2006), the astrophysicist, astrobiologist and emerging telegenic spokesperson for post-Space Age science and humanism, first introduced Lovelock's work to Margulis in a letter dated 16 June 1970, sent from Cornell. By then the two had broken up, but they were still in correspondence (my father addressed her as Dr Lynn Sagan after they split but then Mrs Lynn Margulis after she remarried). She had asked him whether he knew anyone working on the composition of Earth's early atmosphere. He brought Margulis and Lovelock into contact with each other, serving as a scientific matchmaker for the development of Gaia theory, which benefitted from the blend of Margulisian microbial biology and Lovelockian atmospheric chemistry, although he remained sceptical of Lovelock's thesis that Earth was a self-regulating planet. It is perhaps ironic that the person to broker the meeting was the most visible and vigorous public spokesperson for the search for life beyond Earth in the late twentieth century.[*]

In 1984, Margulis and I published a co-written article called 'Gaia and Philosophy'.[1] It made a link between the idea of Gaia, whose name is mythological but whose impulse is scientific, and philosophy. Looking back on this essay some forty years on, after the death of its second author, is an opportunity to take stock. Gaia can be considered the greatest success of technological humankind's attempts to find life elsewhere in space. It is the discovery that any *living* planet, insofar as it is like Earth, would be a self-regulating system. And, further, that an actively regulated planetary atmosphere that is not in thermodynamic equilibrium is a strong indication that life may be present at the surface. The contrast with traditional Western geological and geochemical ideas could not be more stark: Earth is no more a rock with some life on it than you are a skeleton infested with cells.

---

[*] In the summer of 1956, Sagan was mooning over his unfaithful girlfriend, Lynn, who had gone off on an anthropology trip to Mexico. Meanwhile, he was at McDonald Observatory in the Fort Davis Mountains in West Texas, at the time jointly run by the University of Chicago and the University of Texas. The stars were beautiful and clear there to the naked eye, and Sagan, who was by no means a womaniser, wrote to Margulis of two women named Maria whom he had been seeing. 'Marias' (dark plains – from Latin for 'seas') are common features on the Moon and Mars, visible on the near side of the moon. I can't be sure but I suspect that one or both Marias were made up in an elaborate game (not without self-deception) of intellectual-emotional retaliation for Margulis' infidelities. But by 1970, remarried and with a child on the way, Sagan was happy to hook her up with Lovelock, the better to explore the discovery of earthly life in a new form, if not extraterrestrial beings among the stars.

3

Lovelock and Sagan shared an office at the Jet Propulsion Laboratory in Pasadena, California in the late 1960s, where Lovelock had been hired by NASA to study the detection of life on Mars. Viking's spectroscopic results of Mars's atmosphere showed that it was mostly carbon dioxide, with only trace amounts of oxygen. This was in striking contrast to Earth, whose atmosphere is very complex and contains many compounds that should not even exist given the ordinary rules of chemical mixing. Comparing the atmospheric compositions of Mars and Earth, Lovelock felt sure no life was on Mars. Joshua Lederberg, one of the lead scientists working at the laboratory in the lead-up to the Viking Mars missions – the goal of which was to investigate life on Mars – coined the term 'exobiology' to describe the scientific search for life beyond Earth. In a letter from the early 1970s, Lovelock told Margulis that what the two of them were doing was not really exo- but esobiology: studies of planetary life focused not on space but on home. The blue planet was so rich in prokaryotes, bacteria and archaea, protists and fungi and photosynthesizers, that it was as worthy of study as outer space.[2]

Lovelock was an inventor of instruments that could detect trace gases, including an electron capture detector that allowed the measurement of human-produced hydrocarbons, such as DDT and PCB toxins, in pesticides and consumer products. Thus the reasoning by which he arrived at the Gaia hypothesis was thermodynamic, considering what accounted for the massive chemical and thermodynamic disequilibrium of Earth's atmosphere. He reasoned that Earth's atmosphere contains gases, both exotic and simple, which react with oxygen. For example, methane reacts with oxygen to form carbon dioxide and water. Yet methane and oxygen remain in the atmosphere. What keeps them there? Living beings, Lovelock surmised, used the atmosphere as a sort of external circulatory system, and their activities were observable at a planetary level, as they created measurable quantities of reactive gases in Earth's atmosphere. If life existed elsewhere, organisms would do the same in alien biospheres. The Martian atmosphere, over 95 per cent carbon dioxide, was in chemical equilibrium; its current atmosphere was the end-product of chemical reactions, like the ashes of a fire. There was no need to go there, he told his life-seeking colleagues and NASA bosses, mischievously saying they could save a lot of money: Mars never had life, or if it did, was now a dead planet.

The Gaian commentator Bruno Latour compared Lovelock's electron capture device to a telescope pointed at Earth, showing Earth's surface to belong to an organised, living thing. But Latour, with a Judeo-Christian tinge, further suggested that the discovery of Gaia represents a reversal of the Galilean and Copernican notion of the open universe of many planets, toward a new perspective that recognises Earth as special. However, exactly the opposite is the case. Earth's life, and life in general, is a form of nonequilibrium system which, like other such nonliving complex systems, reduces ambient gradients (differences across a distance), cycles matter internally and maintains form, while producing entropy (a measure of the spread of energy) in exemplary conformity with thermodynamics' second law. Although life on Earth, or rather, the life of Earth, gives its character to the entire planet – making it distinct from other planets so far examined by astronomers – the living process it exhibits, far from being unique, is that of all open thermodynamic systems that feed on available energy and produce wastes – ultimately heat. Due to the aggregate sensing and complex chemical activities of its inhabitants, Earth's complex thermodynamic system[3] exhibits natural purpose[4] – even a kind of mind.

The implicit mindful and purposive nature of open thermodynamic and cellular systems requiring energy is displayed, for example, by sunflowers bending toward the sun as they photosynthesise throughout the day, in bacteria swimming up a sugar gradient, and anaerobic microbes using Earth's magnetic field to escape the surface oxygen which poisons them. *Physarum polychephalum,* a plasmodial slime mould (despite its name, it's not a fungus but a protoctist) affectionately called 'the blob', has been shown to have memory, to be teachable, and to be able to find the quickest route between multiple points. Purposive behaviour is seen even in very simple arrangements of matter, such as currents of hot air in a cold house, streaming through cracks and keyholes to come to equilibrium with their surroundings. Another example is seen in long-lasting chemical reactions called Belousov-Zhabotinsky or B–Z reactions, which slowly evolve colourful autocatalytic reactions that form striking spirals. The patterns, which are achieved by the reduction of electron potential gradients, depend on temperature, pressure and chemical concentration. The oscillations can produce wave fronts that change colour from yellow to clear, clear to yellow, and so on, several times a minute. At Georgetown University, these patterns caught the eye of a Jesuit priest, who asked their carrier, Professor Bill Early, if they were alive. 'No, Father,' said the Professor, 'they are like you: they metabolise but do not reproduce.' These examples suggest not only the fact that natural life is a complex thermodynamic system, but that material systems exhibit intelligence in extracting energy via cycles. Do we want to claim exclusive possession of mind and purpose when the very cells of our brains have multiple precursors in a far more-than-human natural world? The robust and complex relationship between life and thermodynamic systems de-anthropomorphises the question of intelligence without granting it to machine systems, which have no innerness or sensation apart from their human connections.

Cybernetic feedback can be said to exist in thermodynamic gradients used by systems (living and non-) that grow and retreat as they maintain their complexity in response to available energy. This behaviour was modelled in Lovelock and Andrew Watson's relatively simple 'Daisy World' model, which showed that black and white daisies with their varying albedo (reflectivity) – the dark ones tending to absorb and warm, the white ones tending to reflect and cool – could, in principle, steadily cool a planet faced with an increasingly bright sun. A typical example of living cybernetic feedback is the cooling and rainmaking skills of *Emiliania huxleyi*. The ornate marine protist, one of thousands of different kinds of plankton, grows in massive blooms that show up as giant subsea clouds in satellite images. The Venetian blind-like openings of the photosynthetic microbe's ornate exoskeleton let light in and gases out, including the gas dimethyl sulphide, a chemical said to make the sea smell like the sea, and a compound on which water vapour condenses, forming clouds that reflect light and cooling rain. This is one of the many ways that life, by growing and dispersing energy and matter, can regulate its temperature, and other variables, first locally and then on the planetary level.[*]

Gaian feedback is quite complex, involving a sort of hyper-mingling between organisms and environments that over time turns environments into bodies.[5] Gaia itself may not be best described as a reactive, cybernetic system, but rather an anticipative, autopoietic one.[6] Autopoiesis (auto: 'self'; poiesis: 'making') refers to a system – living matter – that is self-reflexive, self-oriented, literally self-producing. True examples of autopoiesis involve metabolism, meaning any of the various ways in which evolved organisms are able to remain alive and reproduce through access to energy and

---

[*] On land, forests are crucial actors. Plants don't turn most of the solar energy they take in into growth. They use it to turn out clouds, putting the rain in rainforests. Evapotranspiration brings water up from the ground to the undersides of leaves. This spreads heat, which is the basic process – energy delocalisation – underscored by the second law. The spread of latent heat via evapotranspiration from plants equals about six tons of dynamite per acre each summer.

material substrate.* The composition of matter is not what defines life: a dead bull in a bullring may consist of the same matter as it did a moment ago when it was alive but is no longer autopoietic. Rather, it is the process by which that matter is cyclically organised to maintain and reproduce: both activities expand the realm of gradient reduction, while reproduction allows for natural selection. Living cells are autopoietic, continuously making themselves as they use energy and create wastes, perchance to be recycled by their brethren. Autopoietic systems are self-oriented: they have an inside and maintain themselves. Autopoietic Gaian global life has arguably made the planetary surface itself autopoietic, continuously producing itself as a living body in cosmic space. Rather than the usual exobiological NASA mantra that Earth's water made it suitable for life, the opposite may be the case: water may have remained on Earth due to the continuous activities of life, which is mostly water.[7] (This is an example of what astrophysicist Erich Jantsch termed 'time-binding':

---

* Remarkably, considering our anthropocentrism, all known forms of metabolism used to get energy and nutrients from different sources already evolved in prokaryotes; but only one of them, aerobic respiration, is exploited for all familiar animals, including humans. Other forms of metabolism include methanogenesis (in archaea), photosynthesis (in cyanobacteria, etc.), anaerobic respiration, sulphate reduction, sulphur oxidation (for example by the white bacterium, *Beggiatoa*), and fermentation. (Although in intense muscular activity, animal cells may revert to anaerobic respiration for cell energy, while grass- and wood-eating animals such as cows and termites incorporate the methanogenic metabolism of their internal microbes.)

the preservation of previous chemical environments by the continuous activities of life.) Geology itself may also be under life's grip, in part due to calcium carbonate produced by the microskeletons of marine protists that undergo apoptosis – programmed cell death – when they become too populous, falling in submarine rains to the bottom of the sea, perchance to be subducted and serve as a lubricant for crashing tectonic plates and subsequent mountain building.[8]

While the natural selection of species contributes to Gaian complexity, it is not essential for autopoiesis. Global autopoiesis is more a product of the ancient metabolic versatility of prokaryotes, including extremophiles capable of spaceship-less space travel. The biochemical virtuosity of gene-trading bacteria and archaea growing and interacting in geological concert seems to be the heart of Gaian planetary operations. And this 'alien' microbial intelligence, likely the ultimate basis of human intelligence (deeply dependent as we are upon neurons and brain cells), is already evident in our cell ancestors, themselves displaying versions of the mind-like behaviour of non-living, non-algorithmic complex systems. While it is certainly possible that Earth's life originated on Earth, perhaps from chemical gradients near submarine vents, many amino acids and almost all of the nucleobases of DNA have now been found in space. Francis Crick, co-author of the 1953 paper unveiling the structure of DNA, and astronomer Fred Hoyle are among those who have argued that the chances of life evolving extraterrestrially and coming to Earth in the form of hardy bacteria holed up in space debris is greater than the chances of life beginning on Earth. Hoyle even went so far as to suggest that Mars's rusty regolith was the result of oxidation of iron by iron-oxidising bacteria; and that many outbreaks of illness, such as the common cold, may initiate from viruses coming from space, especially when the sun's magnetic field wanes, which it periodically does.[9]

The famous Urey-Miller experiments of 1952 showed that atmospheric gases exposed to an energy source naturally produce amino acids. The findings resembled the medieval idea of spontaneous generation: that life can occur from non-life on Earth's surface. One might speculate that the exciting retrieval of a thoroughly discarded idea – mice don't come from rags, nor maggots from meat, but life as we know it comes from the natural chemistry of the early Earth – made a still older ancient idea, the coming of life from space, seem relatively anachronistic. But the pre-Socratic philosopher Anaxagoras's idea of life's cosmic origin, the theory of 'panspermia', continues to be updated. The cosmos may be fertile: astral bodies, including planets, may be seeded with life, some of them becoming in essence extraterrestrial Gaias, autopoietic beings run by microbes in space. Considering the difficulties of communicating with members of our own species, let alone those who speak other languages, let alone other life forms on Earth, it may be naive to think we will find anything remotely resembling humans in space. But that does not mean extraterrestrial beings would not be intelligent in their way, or even far more intelligent than humans, even if we do not recognise them.

If any sufficiently advanced technology is indistinguishable from magic, as Arthur C. Clarke suggested, so too any sufficiently evolved civilisation may be indistinguishable from nature, which, as Heraclitus said, loves to hide. Human beings, as Margulis emphasised, are not special, just recent. Earth seems to be dangerously warming, likely because of human industrial activity, but it has seen hotter periods in the past: in the Eocene 50 million years ago, global mean temperatures averaged 13 degrees Fahrenheit higher than now, with palm trees growing and crocodiles swimming in the Arctic Circle.[*]

---

[*] A major part of early Gaian thinking was the idea that planetary biology was able to cool itself in the face of an increasingly luminous sun. The sun has a presumed 40 per cent increase in luminosity since its origin; it naturally increases electromagnetic energy per unit of time. Despite this, our planet seems to have been keeping itself cool, with evidence of liquid water on Earth's surface for thousands of millions of years.

In medieval Arabic mythology, genies or *djinn* were beings made of fire, while angels were made of light. The comparison bears making: human beings, whose growth, industry and civilisation resemble a raging fire that will not last, contrast with the stable biosphere and its luminous sun. Humans are in no way necessary to its function, a point underscored by the deep contrast between the length of time Gaia has been in existence, some 3.8 billion years, and the life span of human civilisation so far, which represents only about 0.0000026 of life's Gaian existence. While anthropoid apes have been around significantly longer than cities and the technical feats we consider emblematic of our sophistication as a life form, our society, which does not look too stable, rides on a biotic surface that has not only remained stable, but metastable, in continuous existence for billions of years.

Despite the havoc we wreak, we remain a temporary speck within the diversity even of present species, in a system that has been persevering without us for billions of years. In what has been dubbed m-SETI or microbial SETI, scientists and others have proposed that life may have evolved more than once, with different forms wafting around on meteoric matter and other safe havens for extremophiles. They suggest that the reason we have not found intelligent life in outer space is because inherently intelligent ensembles of bacteria are 'much more likely to be the dominant form of cosmic intelligence, and the transfer of such intelligence is enabled by the processes of panspermia'.[10] In which case alter-Gaias, most of them remaining, like ours, essentially microbial, have had plenty of time to evolve.

When we look at recent work in the sciences and humanities, especially the attempt to name an epoch – the Anthropocene – after ourselves, we should balk at the attempt, conscious or not, of doubling down on arrogant human self-centeredness. The 'Anthropocene' is an event, not an epoch.[11] Despite our intelligence and apparent ability to peer haltingly into the depths of cosmic origins and life's organisation, we are dispensable to the global life form, with an apparent cosmic age of about one-third the life of the cosmos (four billion years), when dated from the Big Bang (thirteen billion years). As Margulis liked to say at the end of her lectures, quoting a 1950s song: 'Got along without you before I met you/Gonna get along without you now.' When asked at the end of such lectures to apply her evolutionary knowledge to the future of *Homo sapiens* she would remark that the average age of a backboned species in the fossil record is four million years, and that in our case, the evidence of our presence in the Earth would not be grandiose but 'a very thin layer of iron, from the cars'. As she wrote in 1998:

> The Gaia [theory] is not, as many claim, that 'the Earth is a single organism'. Yet the Earth, in the biological sense, has a body sustained by complex physiological processes. Life is a planetary-level phenomenon and Earth's surface has been alive for at least 3,000 million years. To me, the human move to take responsibility for the living Earth is laughable – the rhetoric of the powerless. The planet takes care of us, not we of it. Our self-inflated moral imperative to guide a wayward Earth or heal our sick planet is evidence of our immense capacity for self-delusion. Rather, we need to protect us from ourselves … We need honesty. We need to be freed from our species-specific arrogance. No evidence exists that we are 'chosen', the unique species for which all the others were made. Nor are we the most important one because we are so numerous, powerful and dangerous. Our tenacious illusion of special dispensation belies our true status as upright mammalian weeds … Less a single live entity than a huge set of interacting ecosystems, the Earth as Gaian regulatory physiology transcends all individual organisms. Humans are not the centre of life, nor is any other single species. Humans are not even central to life. We are a recent, rapidly growing part of a single huge ecosystem at Earth's surface.[12]

NOTES

1 Sagan, D. and L. Margulis. 1984. 'Gaia and Philosophy' in L. Rouner (ed) *On Nature* (University of Notre Dame Press)

2 Clarke, B. and S. Dutreuil. 2022. *Writing Gaia: The Scientific Correspondence of James Lovelock and Lynn Margulis* (Cambridge University Press)

3 Schneider, E. D. and D. Sagan. 2005. *Into the Cool: Energy Flow, Thermodynamics, and Life* (University of Chicago Press)

4 Sagan, D. and J. H. Whiteside. 2008. 'Gradient-Reduction Theory: Thermodynamics and the Purpose of Life' in Schneider, S., J. Miller, E. Crist and P. Boston (eds) *Scientists Debate Gaia: The Next Century* (MIT Press)

5 Margulis, L., L. Rico and D. Sagan. 2003. 'Propiocepción: la internalización del afuera' ('Proprioception: When the Environment Becomes the Body') in Rico, L. (ed) *Banquete* (Planetary Publishers)

6 Rubin, S., T. Veloz and P. Maldonado. 2021. 'Beyond Planetary-Scale Feedback Self-Regulation: Gaia as an Autopoietic System' *Biosystems*, 199, 104314.

7 Harding, S. 2022. 'Gaia and the Water of Life' in Clarke, B. and S. Dutreuil (eds) *Writing Gaia: The Scientific Correspondence of James Lovelock and Lynn Margulis* (Cambridge University Press)

8 Margulis, L. and D. Sagan. 2009. 'Microbes Move Mountains' in Earle, S. A. and L. K. Glover (eds) *Ocean: An Illustrated Atlas* (National Geographic Books)

9 See, for example, González-Toril E., J. Martinez-Frias, J.M.G. Gomez, F. Rull and R. Amils. 2005. 'Iron Meteorites Can Support the Growth of Acidophilic Chemolithoautotrophic Microorganisms' *Astrobiology* 5(3): 406–414. Although the consensus scientific view is that Earth life had a local origin, on probabilistic grounds there might be many more chances for hardy life to evolve in space and be transported than evolving in situ.

10 Slijepcevic, P. and C. Wickramasinghe. 2021. 'Reconfiguring SETI in the Microbial Context: Panspermia as a Solution to Fermi's Paradox' *Biosystems* 206, 104441.

11 Gibbard, P., M. Walker, A. Bauer, M. Edgeworth, L. Edwards, E. Ellis, S. Finney et al. 2022. 'The Anthropocene as an Event, not an Epoch' *Journal of Quaternary Science* 37, 3: 395–399.

12 Margulis, L. 1998. *Symbiotic Planet. A New Look at Evolution* (Basic Books)

Gaia and Philosophy

DORION SAGAN AND LYNN MARGULIS

This essay has been slightly edited and updated since its original publication in 1984.

The Gaia hypothesis is a scientific view of life on Earth that represents one aspect of a new biological worldview. In philosophical terms this new worldview is more Aristotelian than Platonic. It is predicated on the earthly factual, not the ideal abstract, but there are some metaphysical connotations. The new biological worldview, and Gaia as a major part of it, embraces the circular logic of life and engineering systems, shunning the Greek-Western heritage of final syllogisms.

Gaia is a theory of the atmosphere and surface sediments of the planet Earth taken as a whole. The Gaia hypothesis in its most general form states that the temperature and composition of the Earth's atmosphere are actively regulated by the sum of life on the planet – the biota. This regulation of the Earth's surface by the biota and for the biota has been in continuous existence since the earliest appearance of widespread life. The assurance of continued global habitability according to the Gaian hypothesis is not a matter merely of chance. The Gaian view of the atmosphere is a radical departure from the former scientific concept that life on Earth is surrounded by and adapts to an essentially static environment. That life interacts with and eventually becomes its own environment; that the atmosphere is an extension of the biosphere in nearly the same sense that the human mind is an extension of DNA; that life interacts with and controls physical attributes of the Earth on a global scale – all these things resonate strongly with the ancient magico-religious sentiment that all is one. On a more practical plane, Gaia holds important implications not only for understanding life's past but for surviving in its far-more-than-human future.

The Gaia hypothesis, presently a concern only for certain interdisciplinarians, may someday provide a basis for a new ecology – and even become a household word. Already it is becoming the basis for a rich new worldview. Let us first examine the scientific basis for the hypothesis and then explore some of the metaphysical implications. Innovated by the atmospheric chemist James Lovelock, supported by microbiologist Lynn Margulis and named by novelist William Golding, the Gaia hypothesis states that the composition of all the reactive gases as well as the temperature of the lower atmosphere have remained relatively constant over aeons. (An aeon is approximately a billion years.) In spite of many external perturbations from the solar system in the last several aeons, the surface of the Earth has remained habitable by many kinds of life. The Gaian idea is that life makes and remakes its own environment to a great extent. Life reacts to global and cosmic crises, such as increasing radiation from the sun or the appearance for the first time of oxygen in the atmosphere, and dynamically responds to ensure its own preservation such that the crises are endured or negated. Both scientifically and philosophically, the Gaia hypothesis provides a clear and important theoretical window for what Lovelock calls 'a new look at life on Earth'.

Astronomers generally agree that the sun's total luminosity (output of energy as light) has increased during the past four billion years. They infer from this that the mean temperature of the surface of the Earth ought to have risen correspondingly. But there is evidence from the fossil record of life that the Earth's temperature has remained relatively stable.[1] The Gaia hypothesis recognises this stability as a property of life on the Earth's surface. We shall see how the hypothesis explains the regulation of temperature as one of many factors whose modulation may be attributed to Gaia. The temperature of the lower atmosphere is steered by life within bounds set by physical factors. With a simple model that applies cybernetic concepts to the growth, behaviour and diversity of populations of living organisms, Lovelock has most recently shown how, in principle, the intrinsic properties of life lead to active regulation of Earth's surface temperature. By examining in some detail the life of a mythical world containing only daisies (about which, more later), even sceptical readers can be convinced that it is theoretically possible for living, growing, responding communities of organisms to exert control over factors concerning their own survival. No unknown conscious forces need be invoked; temperature regulation becomes a consequence of the well-known properties of life's responsiveness and growth. In fact, perhaps the most striking philosophical conclusion is that the cybernetic control of the Earth's surface by unintelligent organisms calls into question the alleged uniqueness of human intelligent consciousness.

In exploring the regulatory properties of living beings, it seems most likely that atmospheric regulation can be attributed to the combined metabolic and growth activities of organisms, especially of microbes. Microbes (or microorganisms) are those living beings seen only with a microscope. They display impressive capabilities for transforming the nitrogen-, sulphur- and carbon-containing gases of the atmosphere.[2] Animals and plants, on the other hand, show few such abilities. All or nearly all chemical transformations present in animals and plants were already widespread in microbes before animals and plants evolved. Until the development of Lovelock's Daisy World (see below), the discussion of control of atmospheric methane (a gas that indirectly affects temperature and is produced only by certain microbes, known as methanogenic bacteria[*]) has provided the most detailed exposition of the maintenance of atmospheric temperature stability.[3] The concentration of water vapour in the air correlates with certain climatic features, including the temperature at the Earth's surface. The details of the relationship between temperature and forest trees, determining the production and transport of huge quantities of water in a process called evapotranspiration, was recently presented by meteorologists in a quantitative model.[4] Although these scientists did not discuss their

---

[*] The methanogens remain prokaryotes but are now classified as archaea, not bacteria.

work in a Gaian context, they have inadvertently provided a further Gaian example. Indeed, as Hutchinson originally recognised when he described the geological consequences of faeces and, as the new ecology book by Botkin and Keller shows,[5] many observations concerning the effects of the biota in maintaining the environment can be reinterpreted in a Gaian context.[6]

How can the gas composition and temperature of the atmosphere be actively regulated by organisms? Although willing to believe that atmospheric methane is of biological origin and that the process of evapotranspiration moves enormous quantities of water from the soil through trees into the atmosphere, several critics have rejected the Gaia hypothesis as such because they fail to see how the temperature and gas composition of an entire planetary surface could be regulated for several billion years by an evolving biota that lacks foresight or planning of any kind.[7,8]

Primarily in response to these critics, Dr Lovelock and his former graduate student Dr Andrew Watson formulated a general model of temperature modulation by the biota, to which they pleasantly refer as 'Daisy World'. Daisy World uses surface temperature rather than gas composition to demonstrate the possible kinds of regulating mechanisms that are consistent with how populations of organisms behave. Daisy World exemplifies the kind of Gaian mechanisms we would expect to find, based as it is on an analogy between cybernetic systems and the growth properties of organisms. In an admittedly simplified fashion, it shows that temperature regulation can emerge as a logical consequence of life's well-known properties. These include potential for exponential growth, and growth rates varying with temperature such that the highest rate occurs at the optimal temperature for each population, decreasing around the optimum until growth is limited by extreme upper and lower temperatures. We will describe the Daisy World in detail shortly.

Some such model, explaining the regulation of surface temperature, is required to explain several observations. For example, the oldest rocks not metamorphosed by high temperatures and pressures, both from the Swaziland System of southern Africa[9,10,11] and from the Warrawoona Formation of western Australia,[12] contain evidence of early life. Both sedimentary sequences are over three billion years old. From three billion years ago until the present, we have a continuous record of life on Earth, implying that the mean surface temperature has reached neither the boiling nor the freezing point of water. Given that an ice age involves less than a 10°C drop in mean mid-latitude temperature and that even ice ages are relatively rare in the fossil record, the mean temperature at the surface of the Earth probably has stayed well within the range of 5° to 25°C during at least the last three billion years. Solar luminosity during the last four billion years is thought by many astronomers to have increased by at least 10 per cent.[13] Thus life on Earth seems to have acted as a global thermostat. Any current estimate for the increase of solar luminosity, which varies from less than 30 to more than 70 per cent,[14] does not alter the outcome of Daisy World's conclusions. A relative increase of solar luminosity from values of 0.6 to 2.2 (its present value is 1.0) is consistent with Daisy World assumptions because a range of values has been plotted by Lovelock and his collaborator Watson.

Cybernetic systems, as is well known to science and engineering, are steered. They actively maintain specified variables at a constant in spite of perturbing influences. Such systems are said to be homeostatic if their variables, such as temperature, direction travelled, pressure, light intensity and so forth, are regulated around a fixed set point. Examples of such set points might be 22°C for a room thermostat or 40 per cent relative humidity for a room humidifier. If the set point itself is not constant but changes with time, it is called an operating point. Systems with operating points rather than set points are said to be homeorhetic rather than homeostatic. Gaian regulatory systems, such as the embryological ones described by C. H. Waddington, are more properly described as homeorhetic rather than homeostatic.[15] Fascinatingly enough, both homeorhetic and homeostatic systems defy the most basic statutes of Western syllogistic thought, although not thought itself, because most people do not think syllogistically but in an associative fashion. For instance, if a person – surely a homeorhetic entity – is hungry, he or she will eat. Thereupon hunger ceases. Put syllogistically, the sense of such a series becomes nullified: I am hungry; therefore I eat; therefore I am not hungry. The thesis leads to an antithesis without ever being synthetically resolved. This circular, tautological mode of operations is characteristic of cybernetic systems, including, of course, all organisms and organismic combinations. It is consonant with the emotive poetic power of contradictory statements, dichotomous personalities and oxymoronic lyrics, such as references to a midnight sun.

Even minimal cybernetic systems have certain defining properties: a sensor, an input, a gain (the amount of amplification in the system), and an output. In order to achieve stability or to increase complexity, the output is compared with the set or operating point so that errors are corrected. Error correction means that the output must in some way feed back to the sensor so that the new input can compensate for the change in output. Positive or negative feedback, usually both, are involved in error correction. A first attempt to apply this sort of cybernetic analysis to the Gaia hypothesis involved development of the Daisy World mathematical model, first by Lovelock and later by Watson and Lovelock together.[16,17] We turn now to the description of the model.

The Daisy World model is used to demonstrate how planetary surface temperature might be regulated. It makes simple assumptions: the world's surface harbours a population of living organisms consisting only of dark and light daisies. These organisms always breed true. Each light daisy produces only light offspring daisies, and each dark daisy produces only its kind. Totally black daisies absorb all of the light coming on them from the sun, and totally white daisies reflect all of the light. The best temperatures for growth for both dark and light daisies are considered to be the same: no growth below 5°C, increasing growth as a function of temperature to an optimum at 20°C and decreasing growth rate above the optimum to 40°C, at which temperature all growth ceases.

At lower temperatures darker daisies are assumed to absorb more heat, and thus to grow more rapidly in their local area than lighter daisies. At higher temperatures lighter daisies reflect and thus lose more heat, leading to a greater rate of growth in their local area. The details have been published in technical journals[18] and have recently been explained in a more popular way by us in the British magazine *The Ecologist: Journal of the Post-Industrial Age*.[19] In summary, the graphs generated by models using these assumptions show that dark and light daisy life can, because of growth and interaction with light, influence the temperature of the planet's surface on a global scale. What is remarkable about the various forms of Lovelock and Watson's model is that the amplification properties of the rapid growth of organisms (here daisies) under changing temperatures are enough in themselves to provide the beginning of a mechanism for global thermal homeorhesis, a phenomenon that some would rather see credited only to a mysterious life force. In general, in these models an increase in diversity of organisms, such as a greater difference between the light and darkness of the daisies, leads to an increase in regulatory ability as well as an increase in total population size.

Daisy World is only a mathematical model. Even with its oversimplification, however, the Daisy World model shows quite clearly that temperature homeorhesis of the biosphere is not something that is too mysterious to have a mechanism. By implication it suggests that other observed anomalies, such as the near-constant salinity of the oceans over vast periods of time and the coexistence of chemically reactive gases in the atmosphere, may have solutions that actively involve life forms. The radical insight delivered by Daisy World is that global homeorhesis is in principle possible without the introduction of any but well-known tenets of biology. The Gaian system does not have to plan in advance or be foresighted in any way in order to show homeorhetic tendencies. A biological system acting cybernetically gives the impression of teleology. If only the results and not the feedback processes were stated, it would look as if the organisms had conspired to ensure their own survival.

The Gaia hypothesis says, in essence, that the entire Earth functions as a massive machine or responsive organism. While many ancient and folk beliefs have often expressed similar sentiments, Lovelock's modern formulation is alluring because it is a modern amalgam of information derived from several different scientific disciplines. Perhaps the strongest single body of evidence for Gaia comes not from the evidence of thermal regulation that is modelled in Daisy World but from Lovelock's own field, atmospheric chemistry.[*]

From a chemical point of view, the atmosphere of the Earth is anomalous. Not only major gases, such as nitrogen, but also minor gases, such as methane, ammonia and carbon dioxide, are present at levels many orders of magnitude greater than they should be on a planet with 20 per cent free oxygen in its atmosphere. It was this persistent overabundance of gases that react with oxygen, persisting in the presence of oxygen, that initially convinced Lovelock when he worked at NASA in the late 1960s and early 1970s that it was not necessary for the Viking spacecraft to go to Mars to see if life was there. Lovelock felt he could tell simply from the Martian atmosphere, an atmosphere consistent with the dicta of equilibrium chemistry, that life did not exist there.[20] The Earth's atmosphere, in fact, is not at all what one would expect from a simple interpolation of the atmospheres of our neighbouring planets, Mars and Venus. Mars and Venus have mostly carbon dioxide in their atmosphere and nearly no free oxygen, while on Earth the major atmospheric component is

[*] Incidentally, Lovelock is an inventor as well as a scientist. He devised the electron capture device, a sensor for gas chromatographs that detects freon and other halogenated compounds in concentrations of far less than one part per million in the air. Indeed, it was Lovelock's invention and observations that in large part sparked off ecological worries of ozone depletion, ultraviolet light-induced cancers and general atmospheric catastrophe.

nitrogen and breathable oxygen comprises a good one fifth of the air.

Lovelock has compared the Earth's atmosphere with life to the way the atmosphere would be without any life on Earth. A lifeless Earth would be cold, engulfed in carbon dioxide and lacking in breathable oxygen. In a chemically stable system we would expect nitrogen and oxygen to react and form large quantities of poisonous nitrogen oxides as well as the soluble nitrate ion. The fact that gases unstable in each other's presence, such as oxygen, nitrogen, hydrogen and methane, are maintained on Earth in huge quantities should persuade all rational thinkers to re-examine the scientific status quo taught in textbooks of a largely passive atmosphere that just happens, on chemical grounds, to contain violently reactive gases in an appropriate concentration for most of life.

In the Gaian theory of the atmosphere, life continually synthesises and removes the gases necessary for its own survival. Life controls the composition of the reactive atmospheric gases. Mars and Venus and the hypothetical dead Earth devoid of life, all have chemically stable atmospheres composed of over 95 per cent carbon dioxide. Earth as we live on it, however, has only 0.03 per cent* of this stable gas in its

---

*The 2022 value is closer to 0.04 per cent.

atmosphere. The anomaly is largely due to one facet of Gaia's operations, namely, the process of photosynthesis. Bacteria, algae and plants continuously remove carbon dioxide from the air via photosynthesis and incorporate the carbon from the gas into solid structures such as limestone reefs and, eventually, animal shells. Much of the carbon in the air as carbon dioxide becomes incorporated into organisms that are eventually buried. The bodies of deceased photosynthetic microbes and plants, as well as of all other living forms that consume photosynthetic organisms, are buried in soil in the form of carbon compounds of various kinds. By using solar energy to turn carbon dioxide into calcium carbonates or organic compounds of living organisms, and then dying, plants, photosynthetic bacteria and algae have trapped and buried the once-atmospheric carbon dioxide, which geochemists agree was the major gas in the Earth's early atmosphere.[*] If not for life, and Gaia's cyclical *modus operandi*, our Earth's atmosphere would be more like those of Venus and Mars. Carbon dioxide would be its major gas even now.

[*] In 2021 the carbon dioxide in Earth's atmosphere was measured to be 414.72 parts per million or .004 percent, or .00041 of Earth's atmosphere, still not much in relative terms when we compare to Mars and Venus, both with atmospheres of over ninety-five per cent carbon dioxide. The rise may be in part due to deforestation. Besides CO_2 production from industry, other factors in recent global warming include pollution by larger particles, for example from gasoline additives and coal burning, which have been shown to trap and reradiate energy as heat, while sometimes also blocking convection currents that provide escape routes for waste energy into space.

Microbes, the first forms of life to evolve, seem in fact to be at the very centre of the Gaian phenomenon. Photosynthetic bacteria were burying carbon and releasing waste oxygen millions of years before the development of plants and animals. Methanogens and some sulphur-transforming bacteria, which do not tolerate any free oxygen, have been involved with the Gaian regulation of atmospheric gases from the very beginning. From a Gaian point of view animals, all of which are covered with and invaded by gas-exchanging microbes, may be simply a convenient way to distribute these microbes more numerously and evenly over the surface of the globe. Animals and even plants are latecomers to the Gaian scene. The earliest communities of organisms that removed atmospheric carbon dioxide on a large scale must have been microbes. In fact, we have a direct record of their activities in the form of fossils. These members of the ancient microbial world constructed complex microbial mats, some of which were preserved as stromatolites, layered rocks whose genesis both now and billions of years ago is due to microbial activities. Although such carbon dioxide-removing communities of microbes still flourish today, they have been supplemented and camouflaged by more conspicuous communities of organisms such as forests and coral reefs.

To maintain temperature and gas composition at liveable values, microbial life reacts to threats in a controlled, seemingly purposeful manner. Gas composition and temperature must have been stable over long periods of time. For instance, if atmospheric oxygen were to decrease only a few percentage points, all animal life dependent on higher concentrations would perish. On the other hand, as Andrew Watson et al. showed, increases in the level of atmospheric oxygen would lead to dangerous forest fires.[21] Small increases of oxygen would lead to forest fires even in soggy rain forests due to ignition by lightning. Thus the quantity of oxygen in the atmosphere must have remained relatively constant since the time that air-breathing animals have been living in forests – which has been over 300 million years. Just as bees and termites control the temperature and humidity of the air in their hives and nests, so the biota somehow controls the concentration of oxygen and other gases in the Earth's atmosphere.

It is this 'somehow' that worries and infuriates some of the more traditional Darwinian biologists. The most serious general problems confronting widespread acceptance of the Gaia hypothesis are the perceived implications of foreknowledge and planning in Gaia's purported abilities to react to impending crisis and to ward off ecological doom. How can the struggling mass of genes inside the cells of organisms at the Earth's surface know, ask these biologists, how to regulate macroconditions like global gas composition and temperature? The molecular biologist W. Ford Doolittle, for example, a man who because of his work is perhaps predisposed towards viewing evolution at smaller rather than larger levels, sees the Gaia hypothesis as untenable, a motherly theory of nature without a mechanism.[22]

Another scientist, the Oxford University evolutionist Richard Dawkins, is even more forceful in his rejection of the theory. Likening it to the BBC Theorem (a pejorative reference to the television documentary notion of nature as wonderful balance and harmony), Dawkins has extreme difficulty in imagining a realistic situation in which the Gaian mechanism for the perpetuation of life as a planetary phenomenon could ever have evolved. Dawkins, author of *The Selfish Gene*, can only conceive of the evolution of planetary homeorhesis in relation to interplanetary selection: 'The universe would have to be full of dead planets whose homeostatic regulation systems had failed, with, dotted around, a handful of successful, well-regulated planets of which Earth is one.'[23]

These sound like forceful arguments, yet if the critics of Gaia cannot accept the notion of a planet as an amorphic, but viable, biological entity, they must have equal if not greater cause to dismiss the origin of life. Surely at one point in the history of the Earth, a single homeostatic bacterial cell existed that did not have to struggle with other cells in order to survive, because there were no other cells. The genesis of the first cell, on Earth or elsewhere, can no more be explained from a strict Darwinian standpoint of competition among selfish individuals than can the present regulation of the atmosphere. While the first cell and the present planet may both be correctly seen as individuals, they are equally alone, and as such they both fall outside the province of modern population genetics.

Nonetheless, Lovelock, a sensitive man with a deep sense of intellectual mischief, has answered his critics with one of their own favourite weapons: mathematical model making in the form of the aforementioned Daisy World.[24] Not believing that the Earth's temperature and gases can be regulated with machine-like precision for billions of years, because organisms cannot possibly plan ahead, Lovelock's critics reject his personification of the planet into a conscious female entity named Gaia. Originally lacking an explicit mechanism and falling outside the major Darwinian paradigm of selfish individualism, it was and still sometimes is difficult for trained evolutionists to refrain from regarding Gaia as the latest deification of Earth by nature nuts. How can an entangled mass of disjointed struggling microbes, they ask, effect global concert of any kind, let alone to such an extent that we are permitted to think about the Earth as a single organism or, better, as something finer, a recycling planetary supra-organism whose main waste is heat exported into space? The answer, of course, is the kind of analysis explored in Daisy World, and one still waits to see how those who accuse Lovelock of conscious mysticism and pop ecology will respond to it in all its mathematical intricacy.

Perhaps the greatest psychological stumbling block in the way of widespread scholarly acceptance of Gaia is the implicit shadow of doubt it throws over the concept of the uniqueness of humanity in nature. Gaia denies the sanctity of human attributes. If intricate planning, for instance, can be mimicked by cunning arrays of subvisible entities, what is so special about *Homo sapiens* and our most prized congenital possession, the human intellect? The Gaian answer to this is probably that nothing is so very special about the human species or mind. Indeed, recent research points suggestively to the possibility that the physical attributes and capacities of the brain may be a special case of symbiosis among modified bacteria.[25]

In real life, as opposed to Daisy World, microbes, not daisies, play the crucial role in the continual production and control of rare and reactive compounds. Microbial growth is also responsible, possibly through the production of heat-retaining gases as well as the changing coloured surfaces, for the continuing thermostasis of the Earth. Evolutionarily, microbes were responsible for the establishment of the Gaian system. Insofar as larger forms of animal and plant life are essentially collections of interacting microbes, Gaia may be thought of as still primarily a microbial phenomenon.[26, 27] We human beings, made of microbes, are part of Gaia no less than our bones, made from the calcium from our cells, are part of ourselves.

In his recent article on classical views of Gaia, J. D. Hughes quoted the ancient Greek work *Economics* by Xenophon: 'Earth is a goddess and teaches justice to those who can learn, for the better she is served, the more good things she gives in return.'[28] In the classical view, that is, of the Greek Gaia or Earth Goddess and the Latin Tellus, the Earth is a vast living organism. The Homeric hymn sings:

Gaia, mother of all, I sing oldest of gods.
Firm of foundation, who feeds all creatures living on earth.
As many as move on the radiant land and swim in the sea

And fly through the air – all these does she feed with her bounty.
Mistress, from you come our fine children and bountiful harvests.
Yours is the power to give mortals life and to take it away.

Although Gaia is reappearing in modern dress, the modern
scientific formulation of the Gaian idea is quite different from
the ancient one. Gaia is not the nurturing mother or fertility
doll of the human race. Rather, human beings, in spite of our
raging anthropocentrism, are relegated to a tiny and unessen-
tial part of the Gaian system. People, like brontosaurus and
grasslands, are merely one of the many weedy components
of an enormous living system dominated by microbes. Gaia
has antecedents not only among the classical poets but even
among scientists, most notably in the work of the Russian V.
I. Vernadsky (1863–1945).[29,30] But Lovelock's Gaia hypothesis
is a modern piece of science: it is subject to observational and
experimental verification and modification.

There is something fresh, new and yet mythologically
appealing about Gaia, however. A scientific theory of an
Earth that in some sense feels and responds is welcome. The
Gaian blending of organisms and environment into one,
wherein the atmosphere is an extension of the biosphere,
is a modern rationalist formulation of an ancient intuitive
sentiment. One implication is that there may be a strong

bio-geological precedent for the time-honoured political and mystical goal of peaceful coexistence and world unity.

Contrary to possible first impressions, however, the Gaia hypothesis, especially in the hands of its innovator, does not protect all the moral sanctions of popular ecology. Lovelock himself is no admirer of most environmentalists. He expresses nothing but disdain for those technological critics he characterises as misanthropes or Luddites, people who are 'more concerned with destructive action than with constructive thought'. He claims, 'If by pollution we mean the dumping of waste matter there is indeed ample evidence that pollution is as natural to Gaia as is breathing to ourselves and most other animals.'[31] We breathe oxygen, originally and essentially a microbial waste product. Lovelock believes that biological toxins are in the main more dangerous than technological ones, and he adds sardonically that they would probably be sold in health food stores if not for their toxicity. Yet there is no clear division between the technological and the biological. In the end, all technological toxins are natural, biological by-products that, though via human beings, are elements in the Gaian system. Similarly, legislation and lobbying attempts, such as the recent furore in the United States over the mismanagement of the Environmental Protection Agency, are nothing more or less than part of Gaian feedback cycles.

Ecologically speaking, the Gaia hypothesis hardly reserves a special place in the pantheon of life for human beings. Recently evolved, and therefore immature in a fundamental Gaian sense, human beings have only recently been integrated into the global biological scene. Our relationship with Gaia is still superficial. Technological intelligence has led to the massive spread of our kind of primate (alien taxonomists would surely regard us as a kind of mutant ape) across Earth's surface, infiltrating multiple ecosystems. Similar to the multiplication of cells in the development of a foetus, our exponential population and industrial increase is not sustainable. A newborn that doubles its weight in its first eight weeks would weigh over eight tons by the age of two. This is not how either multicellular organisms or ecosystems work; organisms that grow too fast run out of food, are destroyed by pathogens, or devoured as a food source: organisms that do not come into concert with their biodiverse ecosystems are destined to perish and humans, despite our temporary 'dominance', are no exception. On the other hand, one can imagine sustainable roles for humans or our descendants. For example, our technological abilities might be honed to prevent a Gaia-cidal cometary impact from destroying all life on Earth. (However, even this seems unlikely as Gaia without humans has recovered from previous major impacts.) Aside from our potential as a nervous early warning system for protecting Gaia, our species might become involved in spreading life to other planets, for example in our own solar system, not just or even at all for human colonisation, as science fiction often imagines, but as a vector for the reproduction of Gaia into extraterrestrial spaces with sufficient solar radiation and water. As Freeman Dyson once suggested, once life gets into outer space (assuming it is not already there) there is no telling what might happen; it may easily kick off its human shoes. Rather than looking at the cosmos as a place to colonise, or worrying about it as a site for potential microbial infection, we might look at it as a place to garden, to foster ecosystems which sustainably recycle their chemical elements. These are the sorts of places, including the global ecosystem of the biosphere, within which we humans ourselves evolved. On the one hand, Gaia was an early and crucial development in the history of life's evolutionary past. Without the Gaian environmental modulating system, life probably would not have persisted. Now, only by comprehending the intricacies of Gaia can we hope to discover how the biota has created and regulated the surface environment of the planet for the last three billion years. On the other hand, the full scientific exploration of Gaian control mechanisms is probably the surest single road leading to the successful implementation of self-supporting living habitats in space, with or without humans. If we are ever to engineer large space stations that replenish their own vital supplies, then we must study the natural technology of Gaia. Still more ambitiously, the terraformation of another planet, for example, Mars, so that it can actually support human beings living out in the open, is a gigantic task and one that becomes thinkable only from the Gaian perspective.

In terms of the metaphysics of inner space, acceptance of the Gaian view leads almost precipitously to a change in philosophical perspective. As just one example, human artefacts, such as machines, pollution and even works of art, are no longer seen as separate from the feedback processes of nature. Recovering from Copernican insult and Darwinian injury, anthropocentrism has been dealt yet another reeling blow by Gaia. This blow, however, should not send us into new depths of disillusion or existential despair. Quite the opposite: we should rejoice in the new truths of our essential belonging, our relative unimportance and our complete dependence upon a biosphere that has always had a life entirely its own.

1 Margulis, L., and J. E. Lovelock. 1974. 'Biological Modulation of the Earth's Atmosphere' *Icarus* 21: 471–489.

2 Ibid.

3 Watson, A. J., J. E. Lovelock and L. Margulis. 1978. 'Methanogenesis, Fires, and the Regulation of Atmospheric Oxygen' *BioSystems* 10: 293–298

4 Shukla, J., and Y. Mintz. 1982. 'Influence of the Land-Surface Evapotranspiration on the Earth's Climate' *Science* 215: 1498–1501

5 Botkin, D. B., and E. A. Keller. 1982. *Environmental Studies: The Earth as a Living Planet* (Charles E. Merrill Pubs)

6 Hutchinson, G. E. 1954. 'Biogeochemistry of Vertebrate Excretion' (American Museum of Natural History)

7 Doolittle, W. F. 1981. 'Is Nature Really Motherly?' *CoEvolution Quarterly* 29: 58–63

8 Garrels, R. M., A. Lerman and F. T. MacKenzie. 1981. 'Controls of Atmospheric Oxygen: Past, Present, and Future' *American Scientist* 61: 306–315

9 Margulis, L. 1982. *Early Life* (Jones and Bartlett)

10 Schopf, J. W. (ed). 1983. *Precambrian Paleobiology Research Group Report* (Princeton University Press)

11 Walter, M. R. 1976. 'Stromatolites: The Main Geological Source of Information on the Evolution of the Early Benthos' in S. Bengtson (ed) *Early Life on Earth* (Columbia University Press)

12 Awramik, S. M., J. W. Schopf and M. R. Walter. 1983. 'The Warrawoona Microfossils' *Precambrian Research* 20

13 Newman, M. J. 1980. 'Evolution of the solar "Constant".' In Ponnamperuma, C. and L. Margulis (eds) *Limits to Life* (Reidel Publishing)

14 Ibid.

15 Waddington, C. H. 1976. 'Concluding remarks' in E. Jantsch and C. H. Waddington (eds) *Evolution and Consciousness* (Addison Wesley)

16 Lovelock, J. E. 1983. 'Gaia as Seen Through the Atmosphere' in P. Westbroek and E. W. de Joeng (eds) *The Fourth International Symposium on Biomineralization* (Reidel Publishing)

17 Watson, A. J., and J. E. Lovelock. 1983. 'Biological Homeostasis of the Global Environment: The Parable of 'Daisy World' *Tellus* 35b: 284–289

18 Ibid. 16, 17

19 Sagan, D., and L. Margulis. 1983. 'The Gaian Perspective of Ecology' *The Ecologist* 13: 160–167

20 Lovelock, J. E., and L. Margulis. 1976. 'Is Mars a Spaceship Too?' *Natural History Magazine* 85: 86–90.

21 Watson, A. J., J. E. Lovelock and L. Margulis. 1978. 'Methanogenesis, Fires, and the Regulation of Atmospheric Oxygen' *BioSystems* 10: 293–298

22 Doolittle, W. F. 1981. 'Is Nature Really Motherly?' *CoEvolution Quarterly* 29: 58–63

23 Dawkins, R. 1982. *The Extended Phenotype: The Gene as the Unit of Expression* (W. H. Freeman & Co)

24   Ibid. 17 and Lovelock, J.E. 1983. Daisy World: A cybernetic proof of the Gaia hypothesis. *CoEvolution Q* 31:66-72

25   Margulis, L., and D. Sagan. 1997. *Microcosmos: Four Billion Years of Evolution from Our Microbial Ancestors* (University of California Press)

26   Kaveski, S., D. C. Mehos and L. Margulis. 1983. 'There's No Such Thing as a One Celled Plant or Animal' *The Science Teacher* 50: 34–36, 41–43

27   Margulis, L., Asikainen, C. A. and Krumbein, W. E. (eds), 2011. *Chimeras and Consciousness: Evolution of the Sensory Self* (MIT Press)

28   Hughes, J. D. 1983. 'Gaia: An Ancient View of Our Planet' *The Ecologist: Journal of the Post Industrial Age* 13: 54–60

29   Vernadsky, V. I., 1998. *The Biosphere* (complete annotated edition: Foreword by Margulis, L. et al., Introduction by Grinevald, J., translated by Langmuir, D. B., revised and annotated by McMenamin, M. A. S.) (Copernicus/Springer)

30   Lapo, A. V. 1988. *Traces of Bygone Biospheres*, translated by V. Purto (Synergistic Press)

31   Lovelock, J. E. 1979. *Gaia: A New Look at Life on Earth* (Oxford University Press)

Images by Anicka Yi

Lynn Margulis (1938–2011) was an interdisciplinary evolutionary biologist, author and educator. She was the primary intellectual force in the twentieth and early twenty-first century responsible for the acceptance of the role of symbiogenesis (of archaea and bacteria) played in the evolution of the eukaryotic cells (cells with nuclei) that became plants, animals, and fungi. She also, by underscoring the roles of gas-exchanging bacteria, helped turn James Lovelock's Gaia hypothesis that the reactive gases of Earth's atmosphere were physiologically regulated by life away from thermodynamic equilibrium, into a full on theory. Her books, often written in collaboration with her son Dorion Sagan, include *Microcosmos: Four Billion Years of Evolution from Our Microbial Ancestors* (1987), *Mystery Dance: On the Evolution of Human Sexuality* (1991) and *Slanted Truths: Essays on Gaia, Symbiosis, and Evolution* (1997).

Writer and ecological philosopher **Dorion Sagan** is author or coauthor of twenty-five books, translated into fifteen languages, including Danish, Japanese, Turkish, Catalan, and Basque, on topics ranging from evolution of the biosphere to the thermodynamics of ecosystems to programmed ageing. He was called an 'unmissable modern master' by *New Scientist*; Nobel laureate chemist Roald Hoffman called his cowritten *Into the Cool* 'fascinating,' and anthropologist Melvin Konner, writing in *The New York Times*, said of his co-authored *Microcosmos* that 'this admiring reader of Lewis Thomas, Carl Sagan and Stephen Jay Gould has seldom, if ever, seen such a luminous prose style in a work of this kind.' His current interests include poetry and experimental literature. With Carl Sagan and Lynn Margulis, his parents, he is co-author of the entries for both 'Life' and 'Extraterrestrial Life' in the *Encyclopedia Britannica*. His current projects include poetry, a story collection, and a 'metabiography' on his parents' early romance and connected sciences.

Anicka Yi (born 1971 in Seoul, South Korea) is a conceptual artist whose work lies at the intersection of fragrance, cuisine and science. She is known for installations that engage the senses, especially the sense of smell, and for her collaborations with biologists and chemists. Yi lives and works in New York City.